MATTRESS SCIENCE FUNDAMENTALS

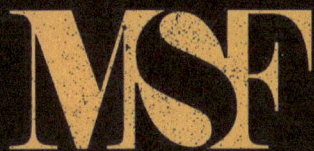

WHERE CREATIVITY MEETS INDUSTRIAL DESIGN

Copyright © 2024 Lérida I Torres Manso

All rights reserved

First Edition

Tampa, Florida
raulyoellafe.escritor@gmail.com

First original publication of Esperanza Editors 2024

ISBN 979-8-3229-2239-1

A BOOK PRINTED IN THE UNITED STATES OF AMERICA

Lérida Inés Torres Manso. Taguayabón, Villaclara, 1957

Graduated in Architecture, Central University of Las Villas, Cuba, 1980

Professor at the Faculty of Construction of the Central University of Las Villas, Cuba, Researcher of Cuban Colonial Architecture General Plan Specialist and head of the Civil Construction Department of the Cienfuegos Oil Refinery. Cuba, 1991. Designed and executed several buildings in that period: Telephone Plant, Administrative Building, Gardening Project, Recreational Club with sauna and swimming pool, Reconstruction of an office building among others. She specializes in construction, investments and development of Hospitality and Tourism. She designed and executed several themed restaurants in the province of Cienfuegos as the Director of Hotels, 2014. Adjunct professor at the University of Cienfuegos, Cuba and at Formatur Cienfuegos, where she also graduated in Hotel Administration, 1998 and in Non-Hotel Administration, 2002. She received and taught numerous postgraduate degrees.

She obtained numerous awards and recognitions for her professional work. She worked on the design, construction, restoration and reconstruction of numerous works in Havana and Cienfuegos, including the Emilio Menéndez Aceval Palace, today known as the Blue Palace, an emblematic building in the city of Cienfuegos. She participated as a speaker at national and international events and conferences inside and outside of Cuba.

She has published several works on Architecture, restoration and Tourism and received awards for several of her works.

Lerida immigrated to Brazil in 2014 and from there, moved to the United States in 2015. She has worked for more than 6 years in the design, manual manufacture and sale of mattresses of different types, specializing in this matter.

MATTRESS Science Fundamentals
A GUIDE TO MATTRESS MANUFACTURING ESSENTIALS

A MESSAGE FROM THE WRITER

WHEN I ARRIVED IN THE UNITED STATES 8 YEARS AGO, I NEVER IMAGINED THAT I WOULD ENTER THE WORLD OF MATTRESS MANUFACTURING AND DESIGN. THANKS TO MARÍA T. CAMINERO FOR INTRODUCING ME TO THE MAGIC OF UPHOLSTERY AND MATTRESS MANUFACTURING AND FOR ALL THE TEACHINGS SHE HAS GIVEN ME IN THIS TECHNICAL SPECIALTY. I AM SO GRATEFUL FOR BEING ABLE TO LIVE IN THIS GREAT NATION SO THAT I AM ABLE TO LIVE MY DREAMS.

SPECIAL THANKS TO MY EDITOR AND BOOK DESIGNER, KATHERINE ST. ROSE AND CARL DE ANGELINE FOR THEIR DEDICATION AND SUPPORT WITH THIS PUBLICATION.

WITH GRATITUDE,
LERIDA TORRES MANSO

INDEX OF CONTENTS
MATTRESS SCIENCE FUNDAMENTALS

- Writer's Biography---2
- A message from the writer-- 3
- Index of contents--4
- The Mattress Manufacturing Process---------------------------------- 5
- The Future of Mattresses-- 12
- Diagram of the components of a metal spring mattress------------13
- The various inner wood components of mattress boxes------------14
- Mattress materials and Their Characteristics-------------------------18
- Mattresses made with Latex Materials---------------------------------20
- Mattresses made with Viscose Elastic Materials--------------------21
- Mattresses made with High Strength Polyurethane------------------22
- The step-by-step Production of Mattresses--------------------------23
- Main Core Technologies--24
- The art of the RV Mattress---28

MATTRESS SCIENCE FUNDAMENTALS

THE MATTRESS MANUFACTURING PROCESS

Book Summary

Through this essay we aim to introduce the reader to the world of mattresses, which is vast and diverse. There are so many types of mattresses, with different materials, dimensions and firmness, that it can be overwhelming to choose the right one. Let me guide you through some key aspects of selecting the best mattress:

1. Factors to take into account when choosing a mattress:
- Personal needs: Consider your build (height and weight), body temperature, favorite sleeping position, number of people who will share the mattress and any lower back conditions.
- Mattress material:

-Memory foam provides an adaptable and ergonomic rest, ideal for relieving lumbar, muscle or joint pain.
-The HR core offers support and firmness depending on the weight of each person.
-Fabrics such as graphene, cool flow and cool graph help regulate temperature and perspiration.
-Pocket spring mattresses offer elasticity and cushioning, especially for couples.
-Latex mattresses are ideal for people with asthma or allergies, since mites cannot live in this material.

2. Firmness:
- Adequate firmness is crucial for spinal health.
- You can find very firm, firm, medium and soft mattresses.
- Consider your body weight and sleeping position when choosing firmness.

Remember that the choice of mattress is personal and should adapt to your specific preferences and needs.

Choosing a mattress is not something that can be taken lightly. If you really want to rest and take care of the health of your back, you cannot only look at the price, but rather that its properties fit with your morphology, habits and way of sleeping. But first, let's remember the rules that you should always keep in mind:

- If you sleep on your side, avoid very hard mattresses
- If you are heavy, choose a mattress with a high firmness.
- If you are hot, pocket spring mattresses are a good option
- If you are allergic to dust and mites, your best option is latex

As you can see, the key to choosing your mattress can be summarized in two ways:
1) the material and
2) its firmness.

MATTRESS SCIENCE FUNDAMENTALS

THE MATTRESS MANUFACTURING PROCESS

MATTRESS PROPERTIES

The properties of each material (or their combination) define whether the mattress is more suitable for a double or single bed, and also how it will adapt to your body.acilisis.

• Memory foam mattresses usually offer great adaptability and a cozy feeling — ideal if you usually sleep on your side or like soft mattresses.

• Pocket spring mattresses offer good bed independence and facilitate air circulation, making them an excellent option for double beds and hot people. In addition, they have a slight bounce that will allow you to turn more easily if you usually move while sleeping.

• Latex mattresses stand out for their flexibility and are ideal for adjustable beds. It is also the only material where mites cannot nest, so it is the best choice if you suffer from asthma or dust allergies.

FIRMNESS

Depending on the position you sleep in and your weight, you will need more or less support.

For example, if you usually sleep on your back or are heavy, you should choose a firmer mattress, while if you are thin or usually sleep on your side, you will appreciate a low or medium firmness that provides good support without being uncomfortable. We hope that our essay is useful to you and that you like it.

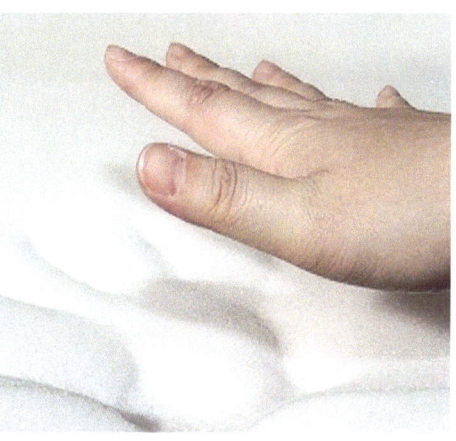

MATTRESS SCIENCE FUNDAMENTALS

THE MATTRESS MANUFACTURING PROCESS

The Mattress Manufacturing Process

The mattress has traveled a long way since the 1850's when uniform springs were used in the mattress upholstery, and the 19th century when the springs were stabilized resulting in a more resistant and firmer mattress.

The mattress industry is on the rise, thanks to the growing demand for a variety of reasons, namely hospitality and health. There is also the booming real estate industry that has led to a growing number of homeowners seeking affordable mattresses that fit their budget. Even the furniture industry is not far behind, since there is also a greater demand for bedding, covers and pillow cases.

With millions of mattresses sold around the world every year, manufacturers continue to find ways to use modern technology to make the mattress manufacturing process more profitable. A great effort was made in the manufacturing of the internal springs of the mattress, since it was considered the core of the bed.

This technical process required a lot of time and the mattress manufacturing subcontracted internal springs or simply bought from a specialized company. In most cases, the real work begins once the internal springs are received. Lately, however, there has been a competition between springs and foam mattresses. The foam mattress, specifically those made with very good quality foam, has gained popularity in recent years thanks to its ability to reduce sleeping points of the sleeping consequences by up to 80%.

MATTRESS SCIENCE FUNDAMENTALS

THE MATTRESS MANUFACTURING PROCESS

Great effort is made to determine the layers of cushion, as it will determine the support or comfort of the mattress. Of course, it is very thought of in the manufacturing process of the outer cover to convert the mattress into a beauty object.

The materials that make up your mattress will determine your sleep quality. You may fall in love with the outside of your mattress at first glance, but later you will know that what matters most is inside the mattress. Whether you want to sleep in the clouds or on a hard surface to get more support, it will be better to know what is inside your mattress. These are some of the mattress materials most used by manufacturers:

Foam: The type of foam used on your mattress will help relieve your pressure points and regulate your temperature. Apart from the viscose-elastic foam that has become popular among those enjoying a repairing sleep, it can also opt for viscose-elastic foam, polyurethane foam and gel viscose-elastic foam, among others.

Polyester Batting: This material is used for filling pillow cover and mattress covers.

Wool: This is a water-resistant insulation that can give your mattress an additional padding.

Cotton: A breathable material can add softness to the mattress.

Steel coils: Provides a good mattress stand.

Quality Control

We will spend at least one third of our lives sleeping or simply resting on our mattress, so it is better to make sure it is safe and has passed quality assurance. Every manufacturer that is precipitated should be ensured that the mattresses only leave the factory once the quality test has passed. This is a standard followed by the mattress industry and should never commit itself not only to keep customers satisfied but also safe.

Before choosing a mattress, make sure that the manufacturer has technicians who check the mattresses for any defect. A company with an established warranty program may minimize defects and, therefore, win more in the long term.

There must be inspections and quality control each stage of the mattress manufacturing.

MATTRESS SCIENCE FUNDAMENTALS

BODY TYPE: Heavier body weights: A firmer mattress is best for heavier bodies because more weight means more pressure on the bed. Too much pressure can cause the bed to sink in and jeopardize spine alignment, leading to back pain. Some top-performing mattress brands also make models designed specifically for people over 250 pounds.

LIGHTWEIGHT SLEEPERS: Smaller frames are better suited with a softer mattress because they aren't putting as much pressure on the bed. If the bed is too firm, it won't sink in enough to relieve pressure on the joints. Factor in both your sleep position and weight when deciding on your best firmness level. For instance, if you're a lightweight stomach sleeper, you can choose a medium firmness to compromise between soft and firm.Sleep Position various positions if you move around at night.

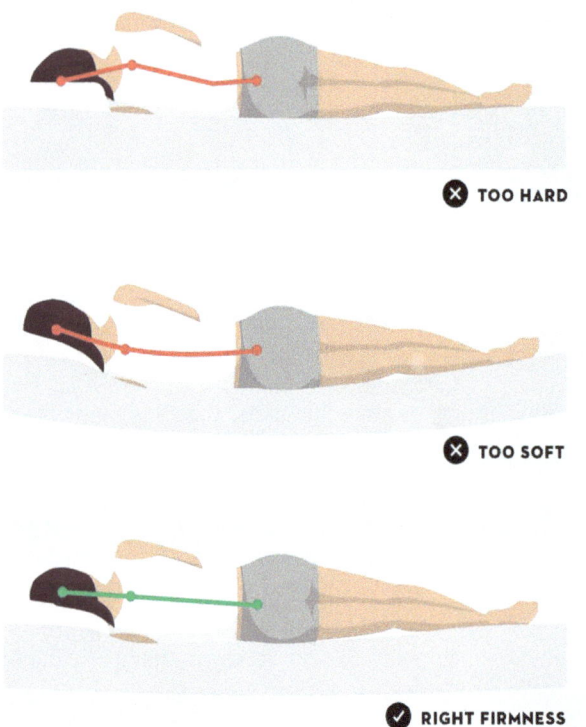

SIDE SLEEPERS: This is the most commonly preferred position and the one that doctors typically recommend to avoid back pain. The best mattresses for side sleepers have soft to medium firmness levels because they help keep your spine aligned. If it's too firm, you may end up putting too much pressure on your hips and shoulders.

STOMACH SLEEPERS: A firmer mattress is more suitable for anyone who likes to sleep on their stomach: You don't want your pressure points to sink in too far in this position.

BACK SLEEPERS: Medium firmness is ideal in this case. If your mattress is too soft or too firm in this position, you risk not having proper alignment.

Combination sleepers: Also opt for medium firmness to best support your various positions if you move around at night.

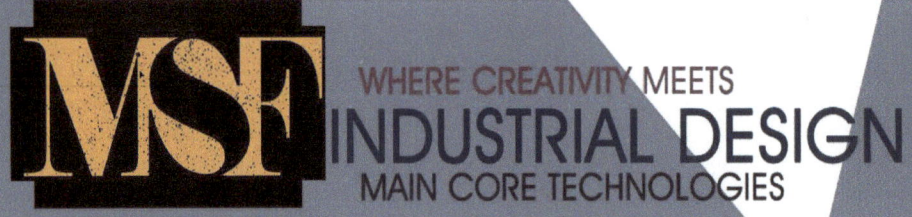

MATTRESS SCIENCE FUNDAMENTALS

SLEEP CONCERNS: Besides creating the right mattress type and firmness level, the manufacturing strattegy should also take into account any specific needs mattress customers have while you shop. Here are common issues and what to look for with each:
HOT SLEEPERS: A cooling mattress can help stay at a comfortable temperature throughout the night, especially for overheating purposes as the mattress customer sleeps. While many factors — like physical conditions and summer heat — can contribute to night sweats and a mattress won't make them magically disappear, the right bed can certainly help alleviate hot sleeping.
Just note that not all cooling materials are the same and memory foam tends to be the worst offender for trapping in heat. Here are common types of cooling features that you'll see when you shop.
BUILT-IN COOLING TECHNOLOGY: Embedded metal particles (like copper), gel and phase-change technology are often used in foam beds to draw heat away from the body. Metal and gel can help prevent overheating, but their cooling effects tend to be less noticeable in real use. Phase-change technology has the ability to store and release heat so it's your best for all-night temperature regulation.

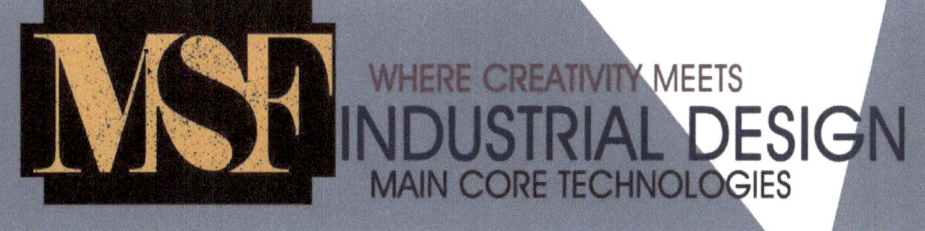

MATTRESS SCIENCE FUNDAMENTALS

COOL-TO-THE-TOUCH MATERIALS: Sometimes you'll notice cooling covers that have an instant chilling effect. These draw in heat immediately, but won't stay cool overnight.
BREATHABLE CONSTRUCTION: Innersprings and some hybrids (with more coils than foam) will allow for more airflow than all-foam mattresses.
ELECTRIC COOLING: There are plug-in options that use water or air to cool the bed. These are ideal to keep your mattress continuously cool, though they're more high maintenance and have added components to incorporate, like a cooling unit next to your bed.
BACK PAIN: The best mattress for someone with back pain will have at least some foam for pressure relief and a medium firmness level for support and spine alignment. We've consulted doctors that specialize in back pain, who say that an underlying issue is likely causing the back pain, but the right mattress can be one step in alleviating discomfort. In fact, studies have shown that the right mattress can improve pain, stiffness and sleep quality up to 50-60%.
ORGANIC MATERIALS: For anyone that prefers an organic mattress made of natural materials, it's important to make sure the entire mattress follows strict organic standards and not just one component. Sometimes brands use an organic cover and call it an organic mattress, which can lead to greenwashing by making it seem more eco-friendly th**an it actually is.**

MSF — WHERE CREATIVITY MEETS INDUSTRIAL DESIGN
MAIN CORE TECHNOLOGIES

MATTRESS SCIENCE FUNDAMENTALS
THE FUTURE OF MATTRESSES

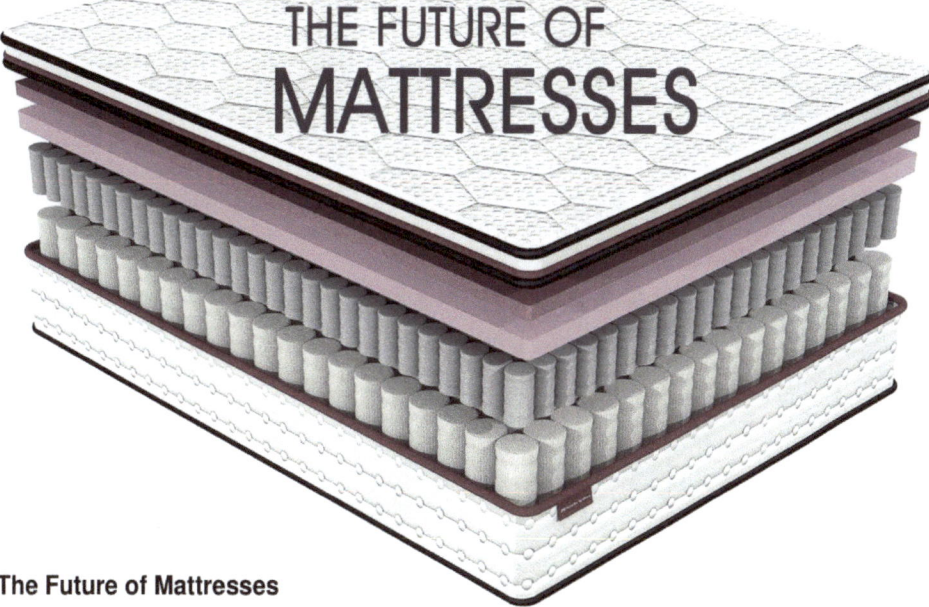

The Future of Mattresses

The mattresses of today are no longer just a place to sleep or rest. The research has demonstrated the important relationship between its sleeping environment and the quality of sleep. Consumers who carry very busy lives no longer take the quality of sleep and are now more critical with the construction of their mattresses.

Apparently, they have discovered that there is an ideal mattress for each and that is no longer a piece of furniture that adapts everyone. The weight, height and preference or position for sleeping from the consumer affect sleep quality. There is a specific mattress that adapts to a person who is overweight, sleeps face up or face down, or someone who sleeps hot.

Despite advances in mattresses technology, manufacturers and sleep scientists have great hurry to find new technologies that help build better mattresses for consumers. This is not surprising at all considering that the world mattress market is expected to be worth more than $ 43 billion over the next five years. Marketing specialists trust that the solid demand for mattresses in China and other emerging and developing countries will result in a better market for the industry.

It is expected that online mattresses will continue to have a great demand, with more than 175 companies in the online mattress sector. The amount of mattresses that are purchased online is also increasing and this was attributed in part to the demographic changes of the buyer. Research shows that Millennials have somehow created a trend in the purchase of mattresses, as they replace their beds more frequently. And as this generation increases its purchasing power, it is expected that mattress manufacturers will take their campaigns to the digital platform.

MATTRESS SCIENCE FUNDAMENTALS
THE MATTRESS MANUFACTURING AND COMPONENTS PROCESS

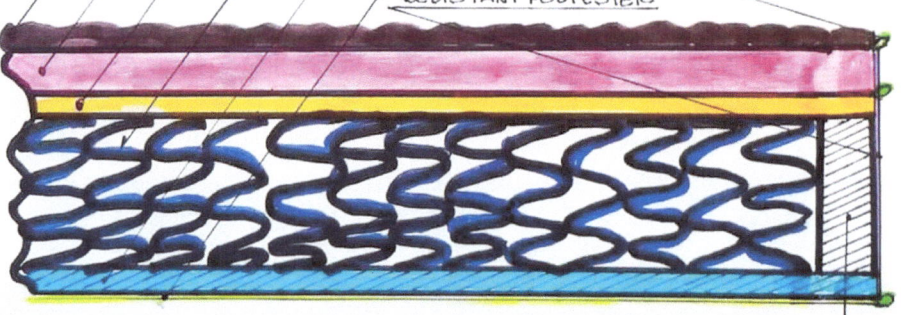

MATTRESS CONSIST OF THE FOLLOWING:

- Topper is 80% POLYESTER and 20% VISCOSE
- 2 INCH FOAM IN PILLOW TOP
- 1 INCH FIBER CORE
- 7 INCH BONNELL SPRING SYSTEM 13 GAUGE INSET
- 1 INCH FOAM CORE BASE
- FIRE RESISTANT COTTON BASE
- TAPE IS 100% FIRE RESISTANT COTTON FIBER
- BORDER IS 100% FIRE RESISTANT POLYESTER
- 3 INCH FOAM ENCASEMENT

PRODUCT DESCRIPTION
AVANT GARDE DESIGN CENTER

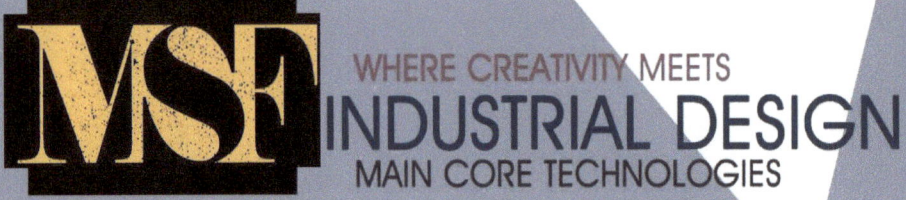

MSF — WHERE CREATIVITY MEETS INDUSTRIAL DESIGN
MAIN CORE TECHNOLOGIES

THE VARIOUS INNER WOOD COMPONENTS OF MATTRESS BOXES

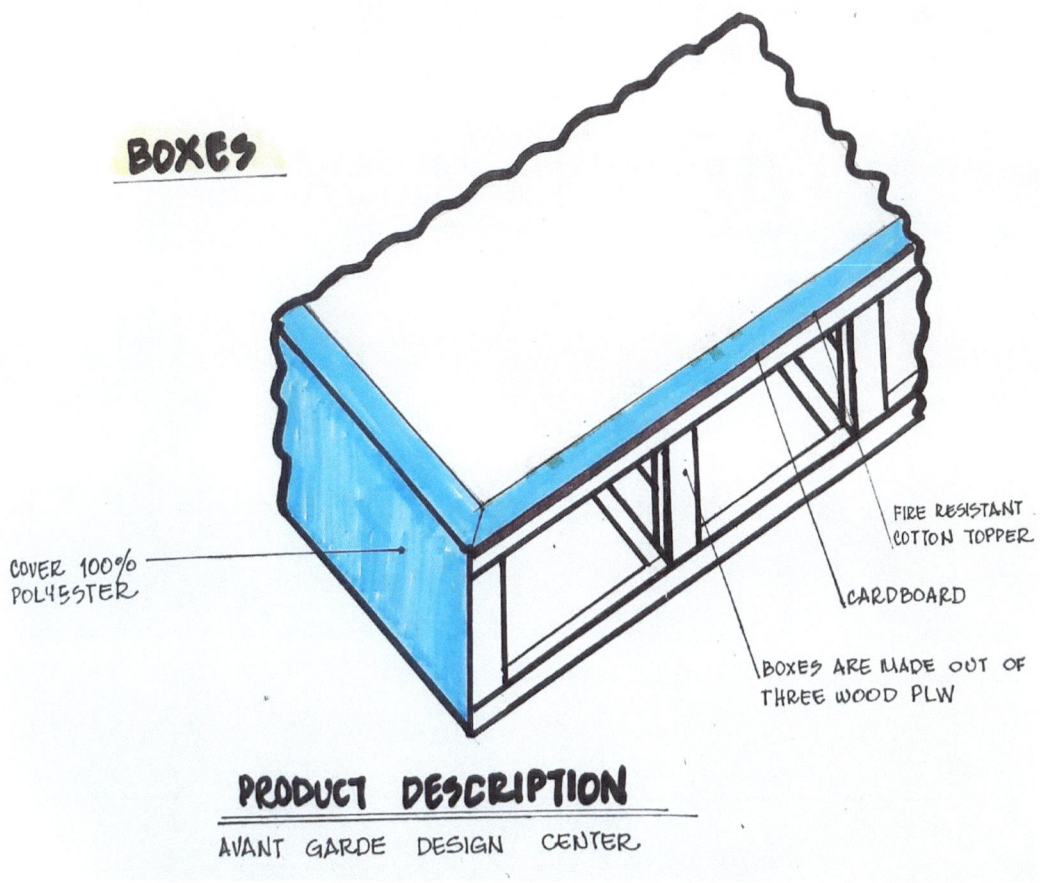

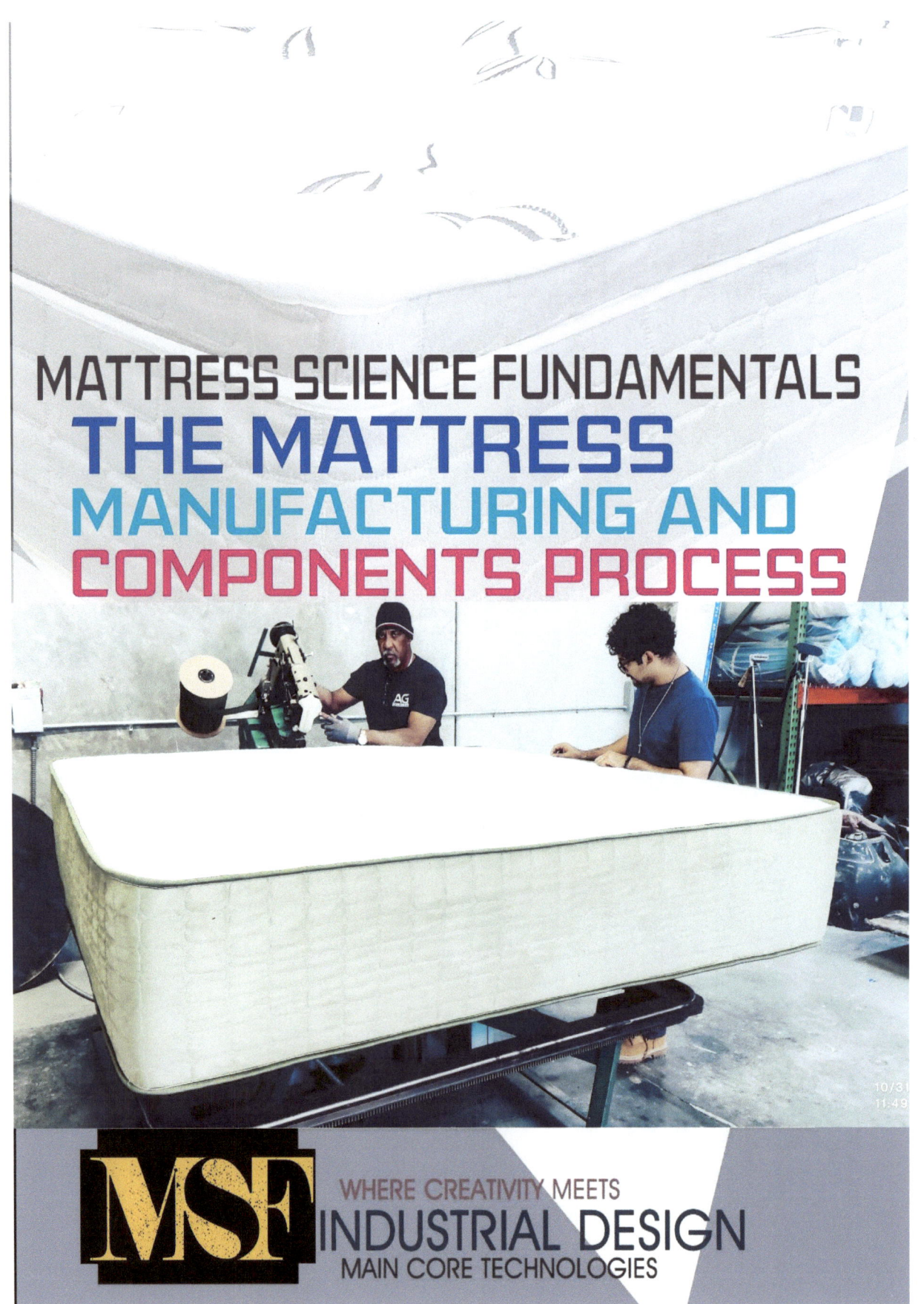

MATTRESS SCIENCE FUNDAMENTALS
THE MATTRESS MANUFACTURING AND COMPONENTS PROCESS

MSF — WHERE CREATIVITY MEETS INDUSTRIAL DESIGN
MAIN CORE TECHNOLOGIES

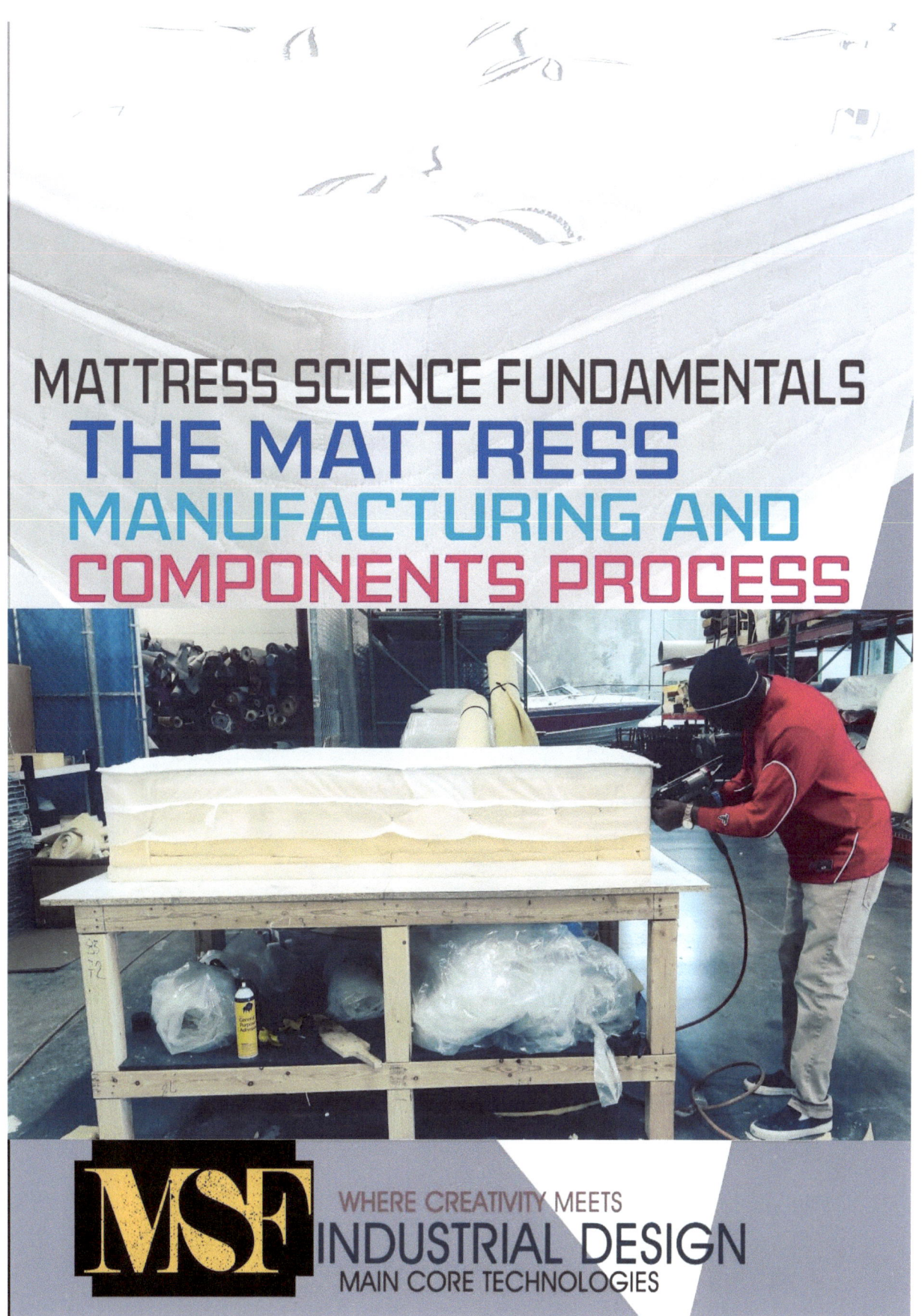

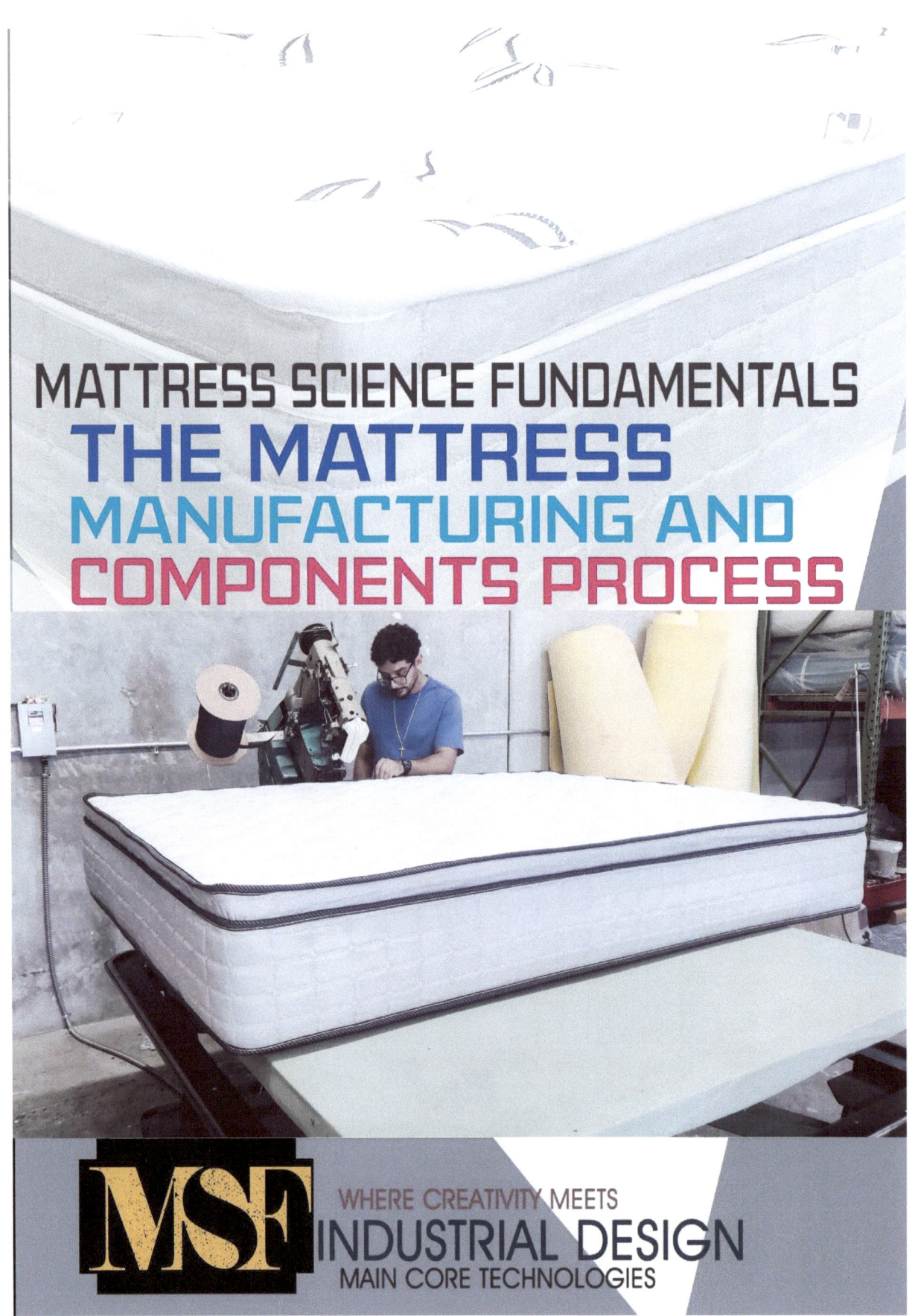

MATTRESS SCIENCE FUNDAMENTALS
THE MATTRESS MANUFACTURING AND COMPONENTS PROCESS

MSF — WHERE CREATIVITY MEETS INDUSTRIAL DESIGN
MAIN CORE TECHNOLOGIES

MATTRESS SCIENCE FUNDAMENTALS
MATTRESS MATERIALS AND
THEIR CHARACTERISTICS

Do you know how many types of mattresses there are on the market?

We will cover the different types of mattresses according to the material with which they are manufactured, as well as the characteristics of each of them so you can choose the one that best suits your preference.

The leading manufacturers in the mattress industry have introduced a large number of technologies and treatments that allow them to market mattresses of great comfort, adaptability, firmness and hygiene. For example, there are treatments that allow the mattress to be free of mites, bacteria and fungi, which can allow you to gain a healthier, more restful sleep.

Before starting with our mattress classification, we will clarify that we are referring to the core material or central block of the mattress, which is the part that provides firmness and the ability to remain sustainable.

The main types of mattress according to their material:

- Viscose-elastic
- Springs
- Espigation or HR
- Mattresses manufactured with springs

MSF
WHERE CREATIVITY MEETS
INDUSTRIAL DESIGN
MATTRESS MATERIALS & CHARACTERISTICS

MATTRESS SCIENCE FUNDAMENTALS
MATTRESS MATERIALS AND
THEIR CHARACTERISTICS

Types of Mattress Springs

Our first type of mattresses refers to spring mattresses. We are talking about the oldest mattresses and are being manufactured since the 19th century.

At present, 3 types of springs are used in the manufacturing mattresses:

- **Independent Springs**

Known as Bonnell or Biconic springs, these hourglass shaped springs adapt to the weight of the different parts of the body. They are put together by steel threads and their structure is not too complicated.

- **Continuous Springs**

Unlike mattresses made with independent springs, this type of core housing is formed by a single continuous thread. This thread is distributed in the form of a "Z" along the entire structure, providing a great consistency to the mattress.

By increasing the density of the wire on the surface of the mattress, the continuous yarn eliminates empty holes, achieving a better substantiation of the body. They offer great firmness and durability, superior to that of biconic mattresses, and this is why their prices are higher.

- **Docking Springs**

These springs have a barrel shape and each of them is introduced into a bag or textile bag to avoid rubbing. Docking springs are especially comfortable and provide an ideal system for double beds. This type of mattress usually has a high durability, thanks to the springs being bagged, which reduces friction, therefore making them more durable over time.

What Types of Mattresses Are More Convenient?

The springed mattress that best suits you depends on your personal preference. What must be taken into account is that the cuffed spring mattress offers you a great independence for both sides of a full, king or queen bed, the dock mattress of continuous thread have great stability and firmness and a bonnelled mattress is firm and economical.

The main advantage of these types of mattresses is their breathability and ability to dissipate heat as well as being easily recyclable.

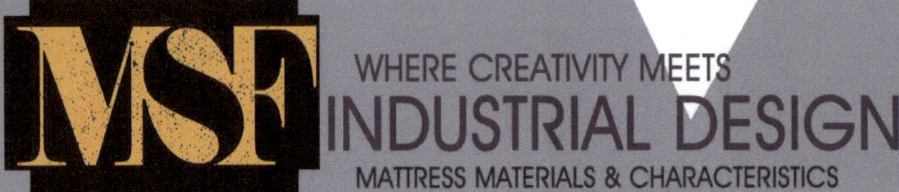

MATTRESS SCIENCE FUNDAMENTALS
MATTRESSES MADE WITH
LATEX MATERIALS

The latex mattresses on the market are of different qualities, depending on the type of latex they contain. Latex is a natural material that is obtained from the sap of a rubber tree or it is synthetically produced from petroleum. In its natural state, latex is a viscous liquid so it is necessary to add chemical additives to adopt a solid state that can be used in the manufacturing of mattresses.

Normally, it is said that a mattress is natural latex when at least 85% of its composition is natural latex. The main advantage of latex mattresses is that they provide comfort and support without adding any other material. They offer adaptability, firmness and are an ideal micro-climate for a restful sleep, which helps with distributing weight evenly.

These mattresses facilitate changes in posture while maintaining a consistent feeling of comfort, which is perfect for those who move a lot at night. Thanks to its flexibility they are ideal for articulated beds.

Like all mattresses with foam cores, perspiration is the weak point of this type of mattress, so this may not be the best choice for those who are susceptible to heat or who perspire in their sleep.

These types of mattresses can be treated to help with respiratory allergies or asthma. However, on the contrary, they should never be used by those who are allergic to latex.

The price of latex mattresses is usually quite high, and depends on the quality, composition and treatments that the mattress has.

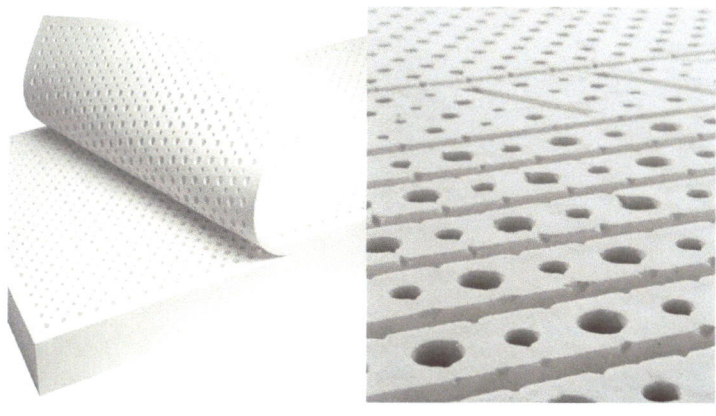

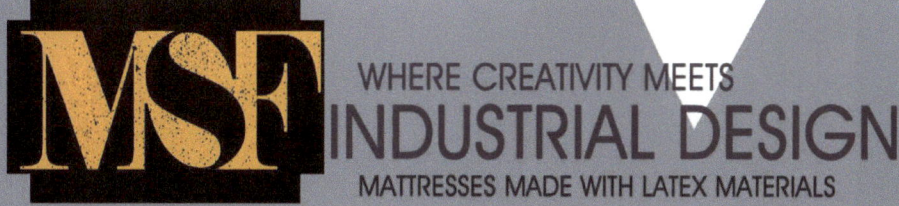

MATTRESS SCIENCE FUNDAMENTALS
MATTRESSES MADE WITH
VISCOSE ELASTIC MATERIALS

These mattresses are made of special, viscose-elastic foam. A material developed by NASA at the end of the 1970's, which, thanks to its adaptable qualities, has become the most used material in the manufacturing of mattresses.

When we talk about viscose-elastic mattresses, we rarely refer to a mattress with a viscose-elastic core, although there are such mattresses, viscose-elastic mattresses do not usually offer sufficient support to be able to be used in the core of a mattress.

The main advantage of this type of mattress is its ability to adapt to the body. Viscose-elastic mattresses react to the weight and temperature of our body, adapting to our body constitution. These types of mattresses are ideal for people looking for a feeling of being cocooned while they sleep.

Viscose-Elastic Mattresses and Their Characteristics

It should be noted that the main handicap of viscose-elastic mattresses is their lack of breathability. They are not suitable for those who are susceptible to heat, since a large part of the body is in continuous contact with the mattress.

There are many types and qualities of viscose-elastic mattresses on the market. From very basic viscose-elastic mattresses with tremendously cheap prices, to others with higher qualities and specific treatments to improve their breathability, durability, and comfort.

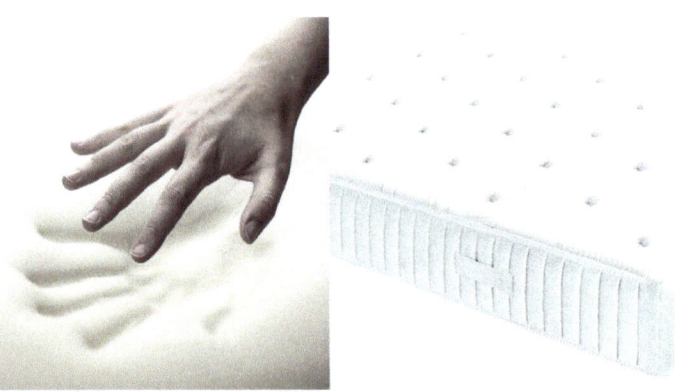

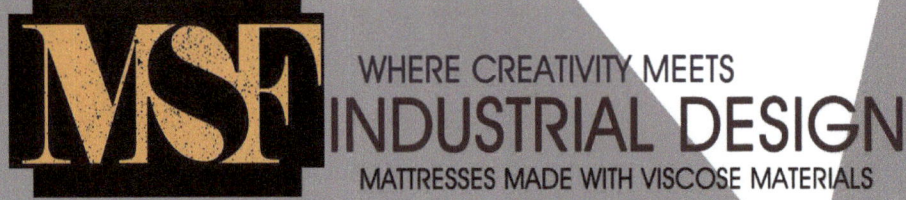

MATTRESS SCIENCE FUNDAMENTALS

MATTRESSES MADE WITH HIGH STRENGTH POLYURETHANE

Mattresses Made with HR or High-Strength Polyurethane

High resilience foam (High Resilience or HR) is a flexible polyurethane alveolar material characterized by an irregular cell structure that offers good flexibility and elasticity. HR is a material that is used in the core of mattresses. Above all, it is used in viscose-elastic mattresses to provide durability to the mattress.

These mattresses offer all the amenities of high density foam and are ideal for second beds. In addition, they have antimicrobial and hypoallergenic properties. They also offer good durability, although they can break down with the passage of time.

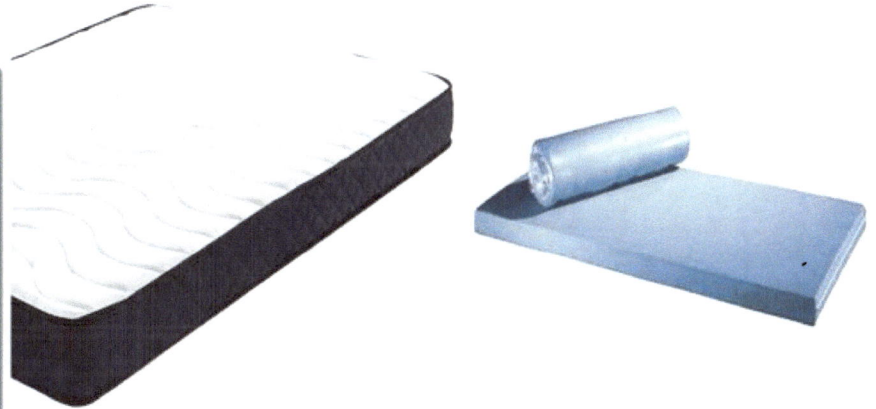

Mattresses Made with Polyurethane

The advantage of this type of mattress is its economic price. As always, they vary depending on density, but generally they are around $200.

The main drawback with polyurethane mattresses is its transpiration capacity, which is aggravated by low quality mattresses. They usually conserve heat and are not very flexible.

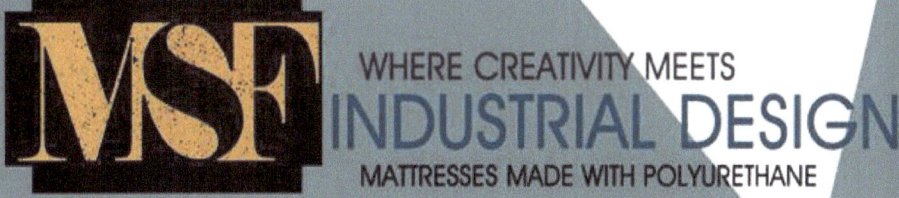

MSF WHERE CREATIVITY MEETS INDUSTRIAL DESIGN
MATTRESSES MADE WITH POLYURETHANE

MATTRESS SCIENCE FUNDAMENTALS
THE STEP-BY-STEP
PRODUCTION OF MATTRESSES

The Step-by-Step Production of Mattresses

For decades, the mattress industry has developed different technologies that combine chemistry and metallurgy to manufacture fascinating polyurethane foams in order to create different sensations from springs and the textile to produce comfort blankets.

There are several different types of mattresses (foam, viscose, latex and springs), and there are also several filling materials. A mattress is actually a set formed by a filling, an upper tissue and sometimes even a general cover. Each layer has its own function, and they are presented in different types. To choose a mattress adapted to the person and that guarantees optimal comfort, it is necessary to discover in more detail the different parts of the mattress.

Each layer of the mattress provides a function:

The soul or core of the mattress determines the support it offers. What determines comfort and in particular the reception (fluffy or firm), are its filling and the upper tissue of the mattress called the mattress cover. The filling and tissue are also determinants for the absorption of moisture or temperature regulations.

MATTRESS SCIENCE FUNDAMENTALS

MAIN CORE TECHNOLOGIES

- Core or foam block

HR

- Core or latex block

- Core or spring block

The Mattress Cover: What Envelops the Mattress

The mattress cover lines of outside of the mattress, and ensures that the different layers of the mattress stay together. Quality and composition are essential to guarantee good sleep and optimal comfort. The mattress cover is the element that is most in contact with the user. It also determines the appearance of the mattress, with various textures and colors.

MSF WHERE CREATIVITY MEETS INDUSTRIAL DESIGN
MAIN CORE TECHNOLOGIES

MATTRESS SCIENCE FUNDAMENTALS

MAIN CORE TECHNOLOGIES

There are plenty of materials to choose from that range from the economical to the highest-range: polypropylene, poly-cotton, and polyester, bamboo viscose-elastic, wool or silk. However, the choice of material is a matter of taste. It is also advisable to check if the mattress cover has been subjected to any extra treatments. It can be an anti-carrier asthma treatment, anti-oil or anti-tier treatment. The nature of the treatment depends on the manufacturer and it may encompass chemical treatments or natural treatments particularly based on essential oils, or probiotics.

The Mattress Filling

The filling of the mattress is responsible for its comfort. According to the materials that compose it, it can be a "fluffy" or "firm" mattress. The filling, however, ensures other functions, in particular the absorption of humidity and temperature regulation. Some innovative materials are also specially designed to help avoid night sweat. For example, foams equipped with vegetable coal thermoregulatory microcapsules, are intended to absorb odors and maintain freshness.

There are many components of natural origin that can ensure a quality mattress, particularly natural latex, one of the three most common technologies. With anti-carrier, antibacterial, breathing and biodegradable properties, natural latex is a material known for its ecological excellence. Other less current materials present the same type of characteristics, such as linen, bamboo, wool, cashmere or silk.

The filling and the upper fabric together determine the comfort of a mattress. The mattress or upper fabric can be formed by different materials, and are chosen according to the tastes of the user and their budget. The mattress cover can be treated with anti-carriers, anti-oil or anti-heat. The technology of the mattress, foam, springs or latex varies according to the properties sought: fluffy or more tonic. The mattresses with 2 faces are covered by 2 types of covers, one ensures more freshness in summer and the other retains coolness in the summer and the other retains heat in winter.

Finally, some manufacturers opt for recycling: some polyester fibers, used for foam mattresses, may be obtained from the recycling of water bottles.

Summer Face and Winter Face

Summer and Winter face are a characteristic of some mattresses as they are made of a different composition on each side. The winter face has a wool composition, which is much better to conserve heat in the winter, while the summer face of the mattress cover is composed of cotton or linen and helps to retain coolness in the summer. To enjoy the characteristics of the different faces, you should turn the mattress regularly.

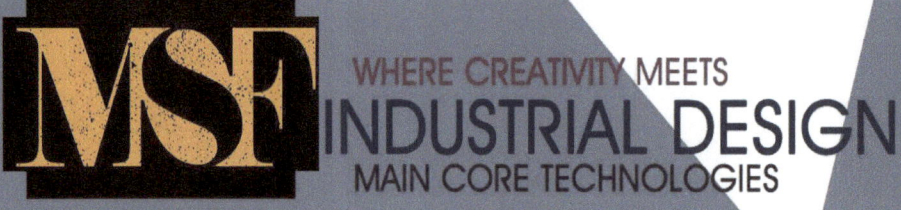

MATTRESS SCIENCE FUNDAMENTALS
MAIN CORE TECHNOLOGIES

Steel wire rolls are used to manufacture springs

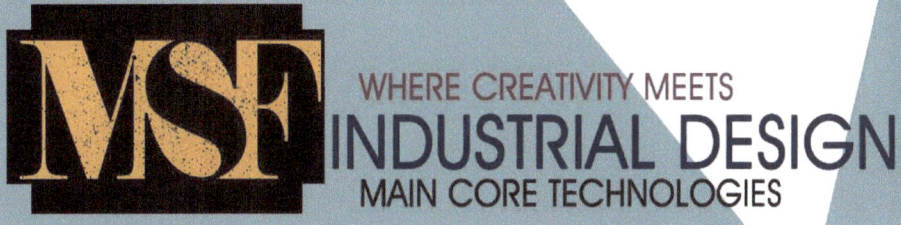

WHERE CREATIVITY MEETS
INDUSTRIAL DESIGN
MAIN CORE TECHNOLOGIES

MATTRESS SCIENCE FUNDAMENTALS

Polyurethane foam is cut according to the thicknesses indicated for each model

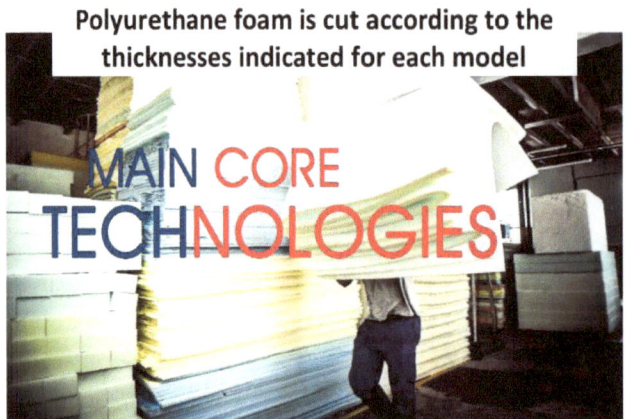

Wire springs are made on machines called resorters

The filling and upper tissue: central elements of comfort

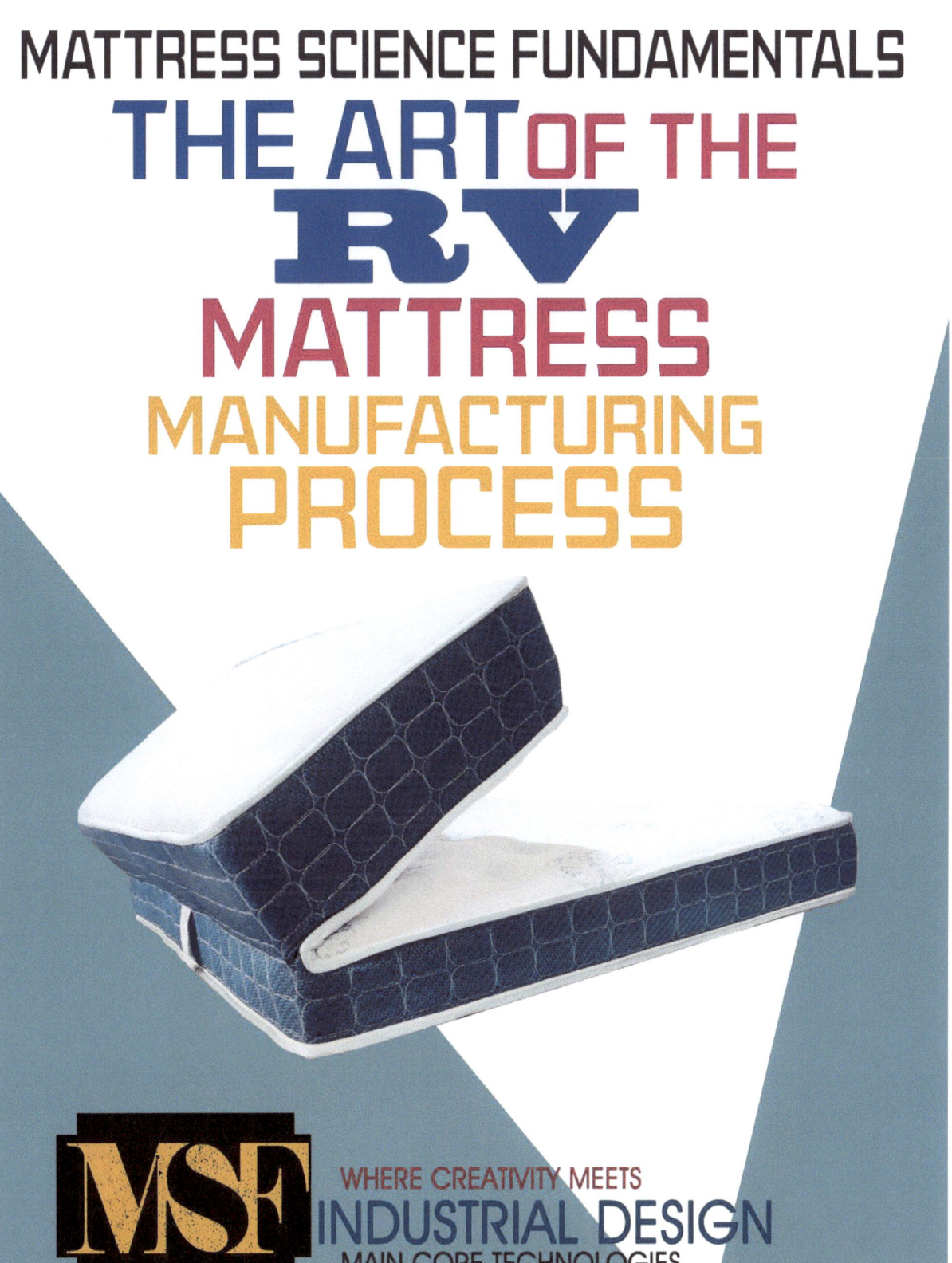

MATTRESS SCIENCE FUNDAMENTALS
THE ART OF THE RV MATTRESS MANUFACTURING PROCESS

WHERE WILL YOUR MATTRESS TAKE YOU?

As an American-Made Mattress brand, we get it. The desire to hit the open road. To satisfy the wanderlust. To venture off the beaten path. Mattress Science Fundamentals creates bedding to be taken on the road less traveled—one of the only companies in the world to design your mattress in our very own state-of-the-art factory in Las Vegas, Nevada. Whatever your road trip aspirations, we promise you the the right RV mattress configuration created and customized for the sleep of your dreams to get there.

YOUR CUSTOM RV AND MARINE MATTRESS DESIGNERS

MSF
WHERE CREATIVITY MEETS INDUSTRIAL DESIGN
MAIN CORE TECHNOLOGIES